BUGS

MOSQUITOES

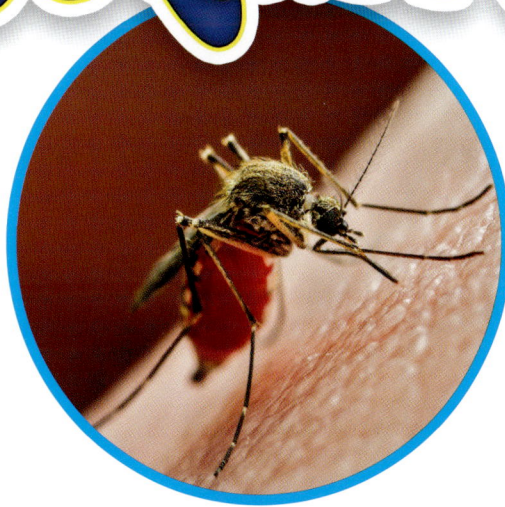

by Emma Huddleston

Consultant: Beth Gambro
Reading Specialist, Yorkville, Illinois

BEARPORT
PUBLISHING

Minneapolis, Minnesota

Teaching Tips

Before Reading

- Look at the cover of the book. Discuss the picture and the title.

- Ask readers to brainstorm a list of what they already know about mosquitoes. What can they expect to see in the book?

- Go on a picture walk, looking through the pictures to discuss vocabulary and make predictions about the text.

During Reading

- Read for purpose. As they are reading, encourage readers to think about the mosquito's life and the impacts the bug has on other things.

- If readers encounter an unknown word, ask them to look at the sounds in the word. Then, ask them to look at the rest of the page. Are there any clues to help them understand?

After Reading

- Encourage readers to pick a buddy and reread the book together.

- Ask readers to name three things from the book that mosquitoes do. Go back and find the pages that tell about these things.

- Ask readers to write or draw something that they learned about mosquitoes.

Credits:
Cover and title page, © nechaev-kon/iStock; 3, © Achkin/Shutterstock; 5, © Aree Saetang/Shutterstock; 6, © moonisblack/iStock, © Pisut chounyoo/Shutterstock; 7, © yod67/Shutterstock; 9, © Papilio/Alamy; 10–11, © Rene Faucher/Shutterstock; 12–13, © Rolf Nussbaumer Photography/Alamy; 15, © khlungcenter/Shutterstock; 17, © Paramonov Alexander/Shutterstock; 19, © blickwinkel /Alamy; 21, © Young Swee Ming/Shutterstock; 22, © jiade/Shutterstock, © thatmacroguy/Shutterstock, © Amir Ridhwan/Shutterstock, © AgriTech/Shutterstock, © srinath82/iStock; 23TL, © Brett_Hondow/iStock; 23TR, © Riska/iStock; 23BL, © koromelena/iStock; 23BM, © Frank Lomagistro/iStock; 23BR, © HAYKIRDI/iStock.

Library of Congress Cataloging-in-Publication Data

Names: Huddleston, Emma, author.
Title: Mosquitoes / by Emma Huddleston.
Description: Bearcub books. | Minneapolis, Minnesota : Bearport Publishing
  Company, 2022. | Series: Bugs | Includes bibliographical references and
  index.
Identifiers: LCCN 2021034182 (print) | LCCN 2021034183 (ebook) | ISBN
  9781636913780 (library binding) | ISBN 9781636913858 (paperback) | ISBN
  9781636913926 (ebook)
Subjects: LCSH: Mosquitoes--Juvenile literature.
Classification: LCC QL536 .H867 2022 (print) | LCC QL536 (ebook) | DDC
  595.77/2--dc23
LC record available at https://lccn.loc.gov/2021034182
LC ebook record available at https://lccn.loc.gov/2021034183

For more information, write to Bearport Publishing, 5357 Penn Avenue South, Minneapolis, MN 55419 Printed in the United States of America.

Contents

Biting Bugs 4

A Mosquito's Life 22

Glossary 23

Index 24

Read More 24

Learn More Online 24

About the Author 24

Biting Bugs

Buzzzz!

A small bug flies by.

It lands on a dog and sucks up blood.

When its belly is full, the mosquito flies away.

There are many kinds of mosquitoes.

A lot of them are brown or orange.

Most are smaller than a paper clip.

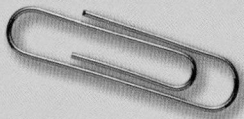

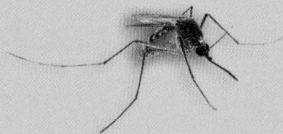

Mosquitoes fly using two thin wings.

Two long **feelers** on their heads help them smell food.

The bugs search for their next meal.

A feeler

Wings

Mouth

Mosquitoes eat the sweet **nectar** from flowers.

They drink with long, tube-shaped mouths.

It is like drinking with a straw.

Slurp!

Mother mosquitoes suck blood, too.

They use it to help make eggs.

Mother mosquitoes can lay 200 eggs at a time.

13

Some mosquitoes make people sick.

They can carry bad **germs**.

The germs **spread** when the bugs bite people.

Mosquitoes can be helpful.

They are food for many animals.

Birds and frogs eat them.

A bat can eat 200 in a single night.

Plants also need mosquitoes.

When they drink from flowers, **pollen** gets on the bugs' bodies.

They spread the pollen and help flowers grow.

19

Most mosquitoes live two to four weeks.

They usually stay in the same area their whole lives.

Buzz, buzz!

A Mosquito's Life

Say pupa like PYOO-puh

Adult

Pupa

Egg

Larva

Say larva like LAHR-vuh

Glossary

feelers two long body parts on a bug's head

germs tiny things that can make people sick

nectar a sweet liquid made in flowers

pollen a dust made by flowers that helps them grow

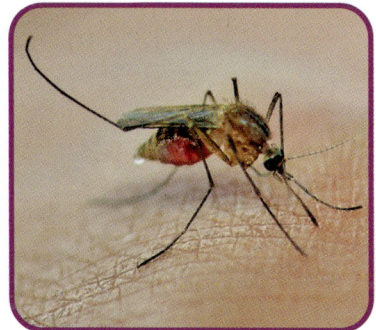

spread to pass from thing to thing

Index

animals 16

blood 4, 12

feelers 8–9

food 8, 16

plants 18

wings 8–9

Read More

Abraham, Anika. *Mosquitoes (Creepy Crawlers).* New York: Cavendish Square, 2019.

Statts, Leo. *Mosquitoes (Insects).* Minneapolis: Abdo Zoom, 2019.

Learn More Online

1. Go to **www.factsurfer.com** or scan the QR code below.
2. Enter "**Mosquito Bug**" into the search box.
3. Click on the cover of this book to see a list of websites.

About the Author

Emma Huddleston lives in the Twin Cities with her husband. She enjoys writing children's books and spending time outside. She often sees the interesting bugs in this series!